NANOTECHNOLOGY

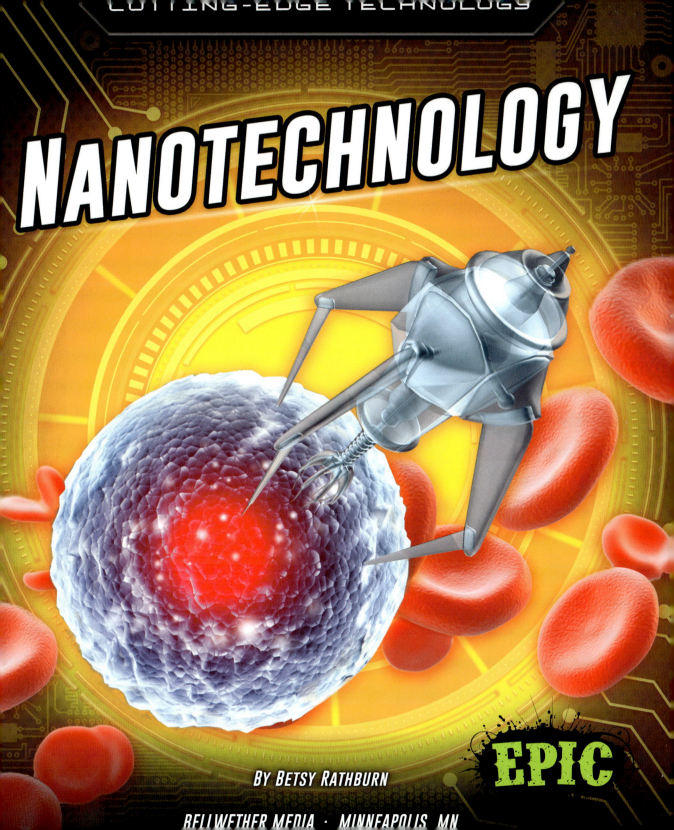

BY BETSY RATHBURN

EPIC

BELLWETHER MEDIA · MINNEAPOLIS, MN

EPIC

Action and adventure collide in EPIC.
Plunge into a universe of powerful
beasts, hair-raising tales, and high-speed
excitement. Astonishing explorations await.
Can you handle it?

This edition first published in 2021 by Bellwether Media, Inc.

No part of this publication may be reproduced in whole or in part without written permission of the publisher.
For information regarding permission, write to Bellwether Media, Inc., Attention: Permissions Department,
6012 Blue Circle Drive, Minnetonka, MN 55343.

Library of Congress Cataloging-in-Publication Data

LC record for Nanotechnology available at https://lccn.loc.gov/2019059305

Editor: Kieran Downs   Designer: Josh Brink

Printed in the United States of America, North Mankato, MN.

TABLE OF CONTENTS

RAINY DAY CLOTHES

You are hiking through the woods. It starts to rain. But your shirt does not get wet. Rain rolls right off!

Your shirt was made with nanotechnology. It kept you dry on your hike!

waterproof fabric

ANOTHER NAME

Nanotechnology is also called nanotech!

WHAT IS NANOTECHNOLOGY?

Nanotech uses tiny materials up to 100 **nanometers** across. Scientists use the materials to make things more useful.

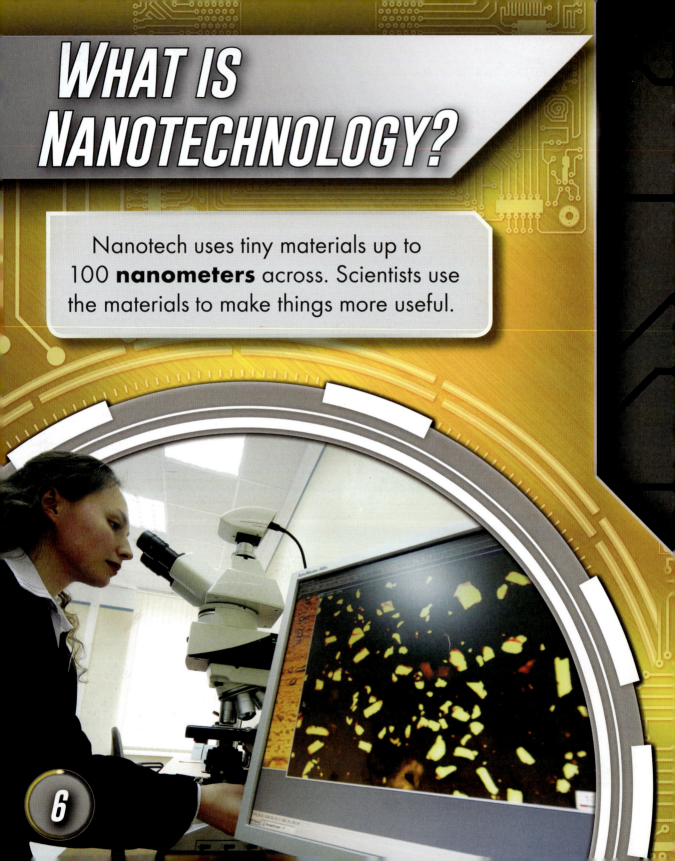

HOW MANY NANOMETERS?

Many tiny objects are thousands of nanometers thick!

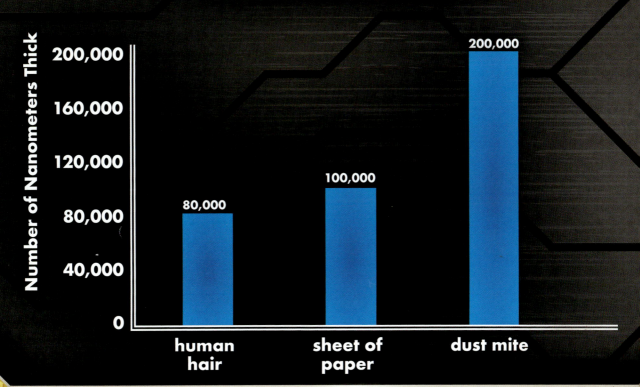

Number of Nanometers Thick

200,000

160,000

120,000

80,000

40,000

0

human hair	sheet of paper	dust mite
80,000	100,000	200,000

Nanotech is all around us. It makes machines lighter. It makes fabric stronger. It even senses changes in health!

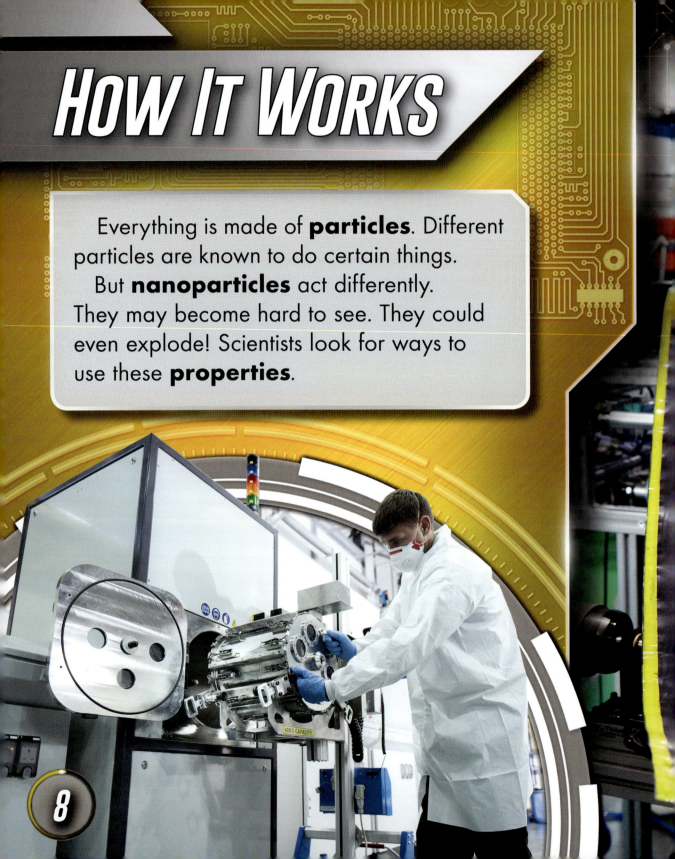

How It Works

Everything is made of **particles**. Different particles are known to do certain things. But **nanoparticles** act differently. They may become hard to see. They could even explode! Scientists look for ways to use these **properties**.

nano material

Microscopes help scientists see materials up close. They help scientists study nanotech. Scientists may cover objects in nanoparticles. This gives the objects new uses!

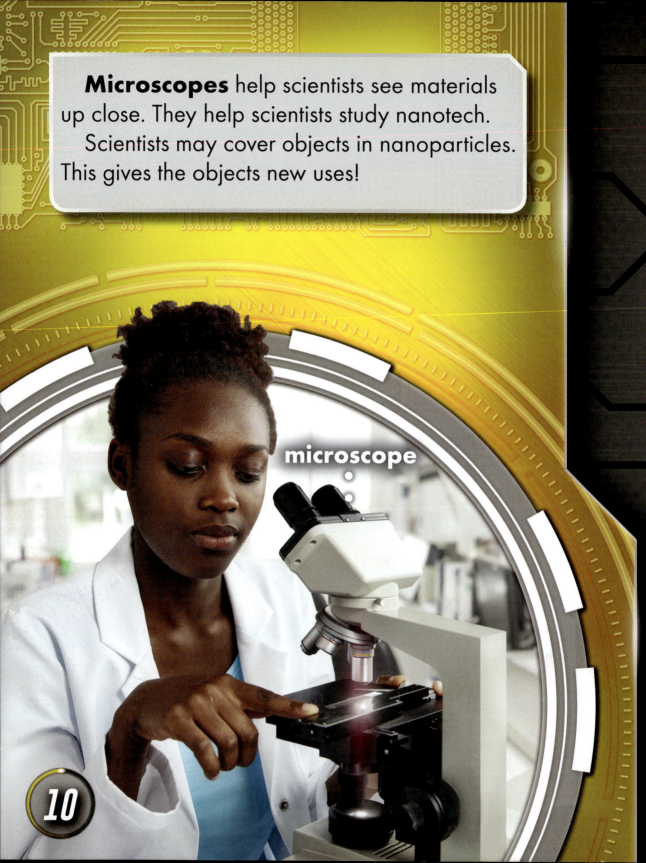

microscope

WATERPROOF FABRIC

Nanotech can be used to waterproof fabric.

nanoparticle coating • • • • •

• • • • • • • water

fabric

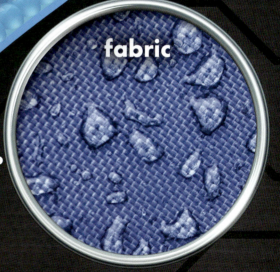

HISTORY

Modern nanotech began in 1959. Richard Feynman spoke about how nanoparticles could be used. But microscopes were not powerful enough.

Richard Feynman

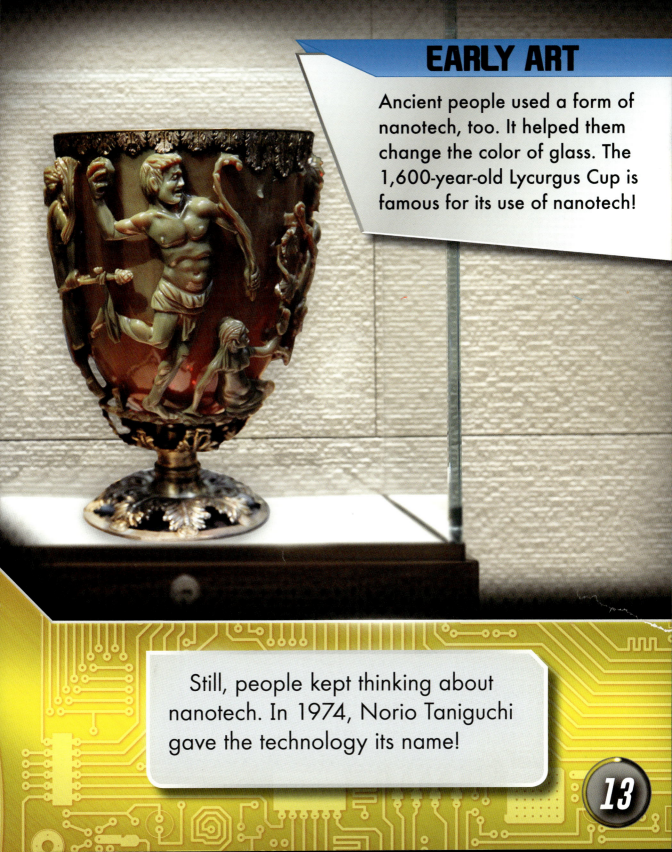

EARLY ART

Ancient people used a form of nanotech, too. It helped them change the color of glass. The 1,600-year-old Lycurgus Cup is famous for its use of nanotech!

Still, people kept thinking about nanotech. In 1974, Norio Taniguchi gave the technology its name!

13

Nanotech Timeline

1959

Richard Feynman speaks about nanotech

1992

New nanotech discovered to help make gasoline

1988

Nanotech used to make new sunscreen

2004

First nanotech bandages heal cuts faster

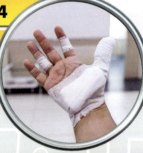

The 1980s brought a more powerful microscope. It helped scientists find uses for nanotech.

2018

New nanotech windows respond to hot and cold

2005

Nanotech baseball bats are released

2019

Scientists announce nanotech that could remove plastic trash from the ocean

2008

New nanotech rubber makes car tires stronger

By the early 2000s, some **consumer** products used nanotech. People could buy many useful items!

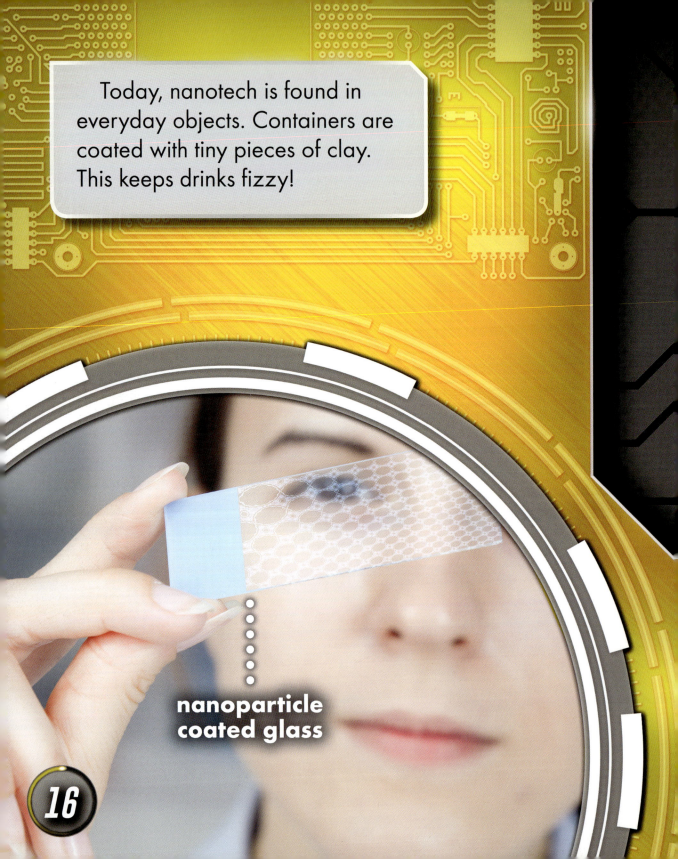

Today, nanotech is found in everyday objects. Containers are coated with tiny pieces of clay. This keeps drinks fizzy!

nanoparticle coated glass

16

WHO USES IT?

military

doctors

scientists

tech companies

Nanocarbon makes metal stronger. Other materials help batteries charge faster!

17

TECHNOLOGY OF TOMORROW

Nanotech could limit **climate change**. Lightweight carbon metal machines use less fuel.

Better **solar panels** could be made with nanotech. Light-reflecting nanoparticles trap more energy!

solar panels

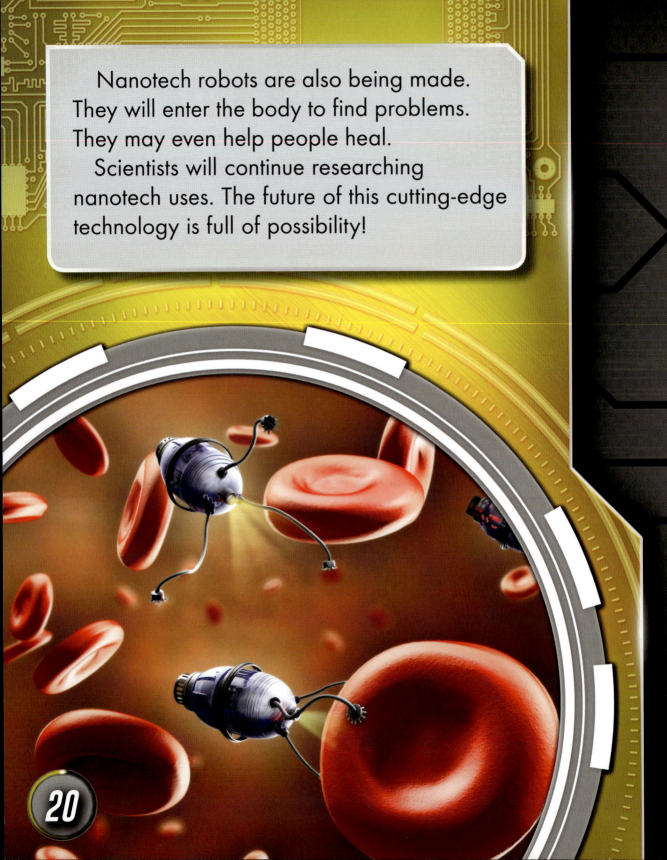

Nanotech robots are also being made. They will enter the body to find problems. They may even help people heal.

Scientists will continue researching nanotech uses. The future of this cutting-edge technology is full of possibility!

PROS AND CONS

Pros

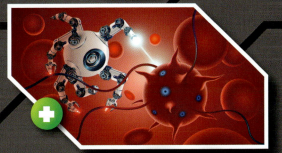

could make people healthier

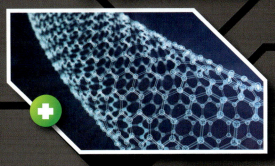

stronger, lighter materials

could limit climate change

Cons

could be used as a weapon

less privacy

expensive

GLOSSARY

climate change—a human-caused process in which Earth's average weather changes over a long period of time

consumer—a person who buys products

microscopes—tools used by scientists to look closely at materials

nanocarbon—a tiny carbon-based material used to make things stronger; carbon is an element found in nature.

nanometers—units of measurement equal to one billionth of a meter

nanoparticles—tiny pieces of materials that can only be seen with microscopes; nanoparticles are measured in nanometers.

particles—tiny pieces of a material

properties—features of a person or thing

solar panels—panels that transform sunlight into energy

TO LEARN MORE

AT THE LIBRARY

Amstutz, Lisa J. *Discover Nanotechnology*. Minneapolis, Minn.: Lerner Publications, 2017.

Gitlin, Marty. *Nanomedicine*. Ann Arbor, Mich.: Cherry Lake Publishing, 2018.

Kulz, George Anthony. *Nanotechnology*. Chicago, Ill.: Norwood House Press, 2018.

ON THE WEB

FACTSURFER

Factsurfer.com gives you a safe, fun way to find more information.

1. Go to www.factsurfer.com.

2. Enter "nanotechnology" into the search box and click 🔍.

3. Select your book cover to see a list of related content.

INDEX